THE AFRICAN MARIJUANA GOLDMINE

RICH MINERAL SOIL MEETS WALL STREET

MICKEY DEE

Frazier Publishing & Services

P.O. Box 363835

North Las Vegas, NV 89036

© Copyright 2018 by Mickey Dee

All rights reserved. No part of this book can be reproduced without the prior permission of the author and publisher.

Disclaimer: This document is geared towards providing exact and reliable information in regards to the topic and issue covered. The publication is sold with the idea that the publisher is not required to render accounting, officially permitted, or otherwise, qualified services.

In no way is it legal to reproduce, duplicate, or transmit any part of this document in either electronic means or in printed format. Recording of this publication is strictly prohibited and any storage of this document is not allowed unless with written permission from the publisher.

The information provided herein is stated to be truthful and consistent, in that any liability, in terms of inattention or otherwise, by any usage or abuse of any policies, processes, or directions contained within is the solitary and utter responsibility of the recipient reader. Under no circumstances will any legal responsibility or blame be held against the publisher for any reparation, damages, or monetary loss due to the information herein, either directly or indirectly.

The trademarks that are used are without any consent, and the publication of the trademark is without permission or backing by the trademark owner. All trademarks and brands within this book are for clarifying purposes only and are the owned by the owners themselves, not affiliated with this document.

The Author and Publisher has strived to be as accurate and complete as possible in the creation of this book, notwithstanding the fact that he does not warrant or represent at any time that the contents within are accurate due to the rapidly changing nature of the Internet. While all attempts have been made to verify information provided in this publication, the Author and Publisher assumes no responsibility for errors, omissions, or contrary interpretation of the subject matter herein.

Any perceived slights of specific persons, peoples, or organizations are unintentional. In practical advice books, like anything else in life, there are no guarantees of results. Readers are cautioned to rely on their own judgment about their individual circumstances and act accordingly.

This book is not intended for use as a source of legal, medical, business, accounting or financial advice. All readers are advised to seek services of competent professionals in the legal, medical, business, accounting, and finance fields.

If you find any value in this book, would you do me the honor of leaving a review on the Amazon page. Thank you for your support. God bless you.

Table of Contents

Ghana Life: The Legacy of Cannabis 1

More Countries in Africa to Legalize Marijuana 6

Cannabis in Africa 12

Health Benefits of Cannabis in Africa 59

Lesotho: Africa's pioneer when it comes to cannabis .. 84

Ways Legal Cannabis Will Change The World 87

Medical marijuana a huge opportunity for Africa 98

Swazi Gold 106

The Marijuana Business Is Really the Real Estate Business 108

Africa, The Next Big Cannabis Market? 117

GHANA LIFE: THE LEGACY OF CANNABIS

Cannabis is widely smoked in Ghana where it is popularly known as 'wee,' probably derived from the English street name 'weed.' According to Henry Bernstein, writing in the Review of African Political Economy of March 1999, the consumption and export of cannabis in Ghana appears to have expanded significantly since the 1960s, and by the 1980s the transit/re-export of cocaine and heroin was also well established. Venezuela is often mentioned as the source of Ghana's cocaine while its transitory heroin comes from the well known Asian sources, much of it collected by traders from India. Historical links with the United Kingdom ensures that a major part of Ghana's re-exports of hard drugs ends up on British streets.

Kumasi has many features that make it a likely location for the headquarters of an international drugs trading network. Its central market at Kejetia is claimed to be the largest in West Africa and has been a major hub for West African trade over several centuries. With this tradition for trading, its people have spread widely around the world creating an international network with strong ethnic links. Kumasi is home to many of Ghana's 6,000 people of Arab ethnicity, known locally as 'Lebanese,' although not all trace their roots back to the Lebanon. Many Lebanese are involved in the import and export trade, exploiting links with Lebanese communities in other countries. One of the largest expatriate Lebanese communities, said to number about 300,000, is found in Venezuela, the source of most of the cocaine transiting Ghana en route to the UK. It is not surprising therefore to find Lebanese names appearing in media reports of Ghanaians arrested for drugs trafficking.

In addition to their expertise in trading, the people of Kumasi possess other skills of great benefit to drugs traders. Media reports of drug seizures by the authorities often express wonder at the ingenious methods of concealment used in efforts to prevent detection. These range from hollow coconuts and balls of kenke (fermented corn dough), through wood carvings and auto spares to stuffed braziers and elaborate hair styles. Suame Magazine in Kumasi, Ghana's largest informal industrial area, is the home of thousands of grassroots workshops and tens of thousands of skilled artisans. Thousands more skilled artisans are employed in villages around Kumasi in traditional crafts such as narrow-loom (Kente) weaving, glass bead making, wood carving and bronze casting. No doubt these skills are called upon from time-to-time to implement the latest schemes of the traffickers.

In a recent novel, an Englishman employed in a Lebanese-owned company in Kumasi brings together a group of Ghanaians and Lebanese to traffic cannabis, cocaine and heroin to the UK. When the principal character, Kwame Mainu, travels to Coventry to study engineering at Warwick University, he realises that several people he knew in Kumasi are involved in a drugs cartel. On vacation back in Ghana he is impressed by the large houses some of his peers are building in Kumasi and he is tempted to join them to share their prosperity. At the same time he is aware that this path could jeopardise his chances of becoming an engineer and playing a key role in the grassroots industrial revolution in Ghana, and it would not be sanctioned by either his wife or his father. In the course of trying to resolve this dilemma, another opportunity presents itself with equally challenging tensions.

MORE COUNTRIES IN AFRICA TO LEGALIZE MARIJUANA

Attention stoners, there is hope for you in Africa after all. Pay attention and learn. Lesotho recently became the first African country to grant a license for medicinal marijuana. The country's Ministry of Health granted South African medical company, Verve Dynamics a license for the cultivation of marijuana for medical and scientific purposes. While there is seemingly nothing much to celebrate about a licence for medicinal marijuana, this is a marked change in tone. One could even call Lesotho's attitude to marijuana progressive and welcome.

In 2007, the United Nations estimated that 38.2 million adults in Africa used marijuana each year yet it was illegal in all countries. This was (and still is) money being channelled to shadow markets yet legalisation would make sure States get a piece of the pie. More than 10,000 tonnes of cannabis are being produced on the continent annually making this a potentially multi-billion-dollar industry. Thankfully, more countries are starting to see the light.

In Swaziland, public officials have seen that they are sitting on gold (albeit green gold). In 2015, Phiwayinkosi Mabuza, Housing and Urban Development Minister argued that the country should legalize cannabis to boost the economy while the National Commissioner of Police called on the government to do a study to establish whether it was desirable to legalise the drug. Mabuza said, "First world countries have decriminalised the growing and use of dagga. We have to be objective and not hysterical when we approach the subject."

Earlier this year, it was reported that a five-member committee of Swazi lawmakers was investigating legalizing the use of cannabis to boost the country's economy. The committee was appointed by Swazi legislators who believe the kingdom could earn $1.63 billion in a year from cannabis thus tripling its domestic product. The House of Assembly was told on the 31st of March that carpets, army uniforms and medicine could be made from cannabis. It remains to be seen what will happen in Swaziland but the kingdom is clearly hard at work. However, Swaziland seems to be focusing on the medicinal utility of marijuana meaning recreational use might not get the greenlight soon. However, do not despair as South Africa has taken the movement a step further.

South Africa's Western Cape High Court earlier this year declared it legal to grow and smoke dagga in one's home. The judgment now awaits confirmation in the Constitutional Court but there is a catch. The import of the judgment is that "it is still illegal to own or use marijuana. However, the ruling means that South Africans can smoke and grow in the safety of their own homes, with a slightly greater peace of mind that they will not face legal ramifications".

The judgment is based on the right to privacy and not some right to smoke marijuana. Finding there was insufficient evidence of the harmfulness of marijuana to justify limiting an individual's right to privacy, the court found the law unconstitutional. Distribution and public use are still illegal. The government said it would appeal the decision and in the meantime, all the smokers can do is hold their breaths and hope for the best. However, in the words of Jeremy Action, an advocate for legalisation, "Things are looking are very positive for some kind of change."

In Malawi, steps have been taken to legalize hemp, a mild species of weed (chamba in Malawi). One of the major proponents of legalising hemp in Malawi, Kadzamira argues that in the future, the country might have to think about legalising weed. He said, "In the future, we have to think about it but for now, it's a very serious topic. It's an emotive issue, and we're trying as much as possible not to mix it with industrial hemp." Malawi is not quite there yet for the recreational user but industrial hemp will most probably be legalised. This is a start of good things!

In Zimbabwe, the government is reportedly considering an application by a Canadian international company to produce cannabis for medical purposes. This could result in the country legalising production and use of marijuana in selected areas. Cabinet member, Obert Mpofu said, "I also laughed and thought they were joking when I received the inquiry but they are serious. This seems to be big business." Indeed, it is and Zimbabwe is better for considering it.

While the recreational smoker might have to wait for a few more years or decades, cannabis is enjoying a renaissance in Africa. History will prove that colonialists passed laws that banned use of cannabis. The laws are clearly racist legacies and have outlived their usefulness. In fact, United States of America first commissioner of the Federal Bureau of Narcotics summarised the whole philosophy behind the demonization of cannabis as that it "makes darkies think they're as good as white men". African countries have no business maintaining racist laws that are backed by nothing but emotive statements of alternative facts and they are realizing it. The great awakening is here!

CANNABIS IN AFRICA

AFRICA

Africa is confirming its role as a production area, a transit territory and a consumer market. South Africa is probably the world's leading producer of marijuana, but most of it is consumed on the domestic market, which also absorbs the output of neighboring countries, notably Lesotho, where cannabis is the main cash crop. At the other end of the continent, Algeria's market for cannabis products is supplied not only in hashish from Morocco but also, recently, in West African marijuana. In some countries, such as Congo, war is to blame for the boom in illicit crops, which have slowly become the prize of rural conflicts. More peacefully, in Kenya, khat represents major political and economic stakes. African and international organized crime are also involved in transit-trafficking of hashish, heroin and Mandrax from South-east Asia, and of South American cocaine. In addition, they distribute these substances on booming African markets. ...

THE CANNABIS INDUSTRY IN LESOTHO

Lesotho is a small, mostly rural, mountainous Southern African country of about 2 million inhabitants, which is completely landlocked in South African territory. Although it is politically an independent state, Lesotho's geographic location makes it very dependent on its powerful neighbor, which absorbs most of its exports. Additionally, given Lesotho's lack of industry, poor soil and general state of underdevelopment (it is one of the world's poorest countries, with a GNP per capita of US $770 in 1995),

South African mines are the largest employer of Basotho workforce.

Lesotho produces large quantities of cannabis, called matekoane in Sesotho, the language spoken in Lesotho. Although there is a domestic consumer market, Lesotho basically grows cannabis in order to supply the large South African market of marijuana. Cannabis production clearly represents one of the country's three main sources of hard currency, the other two being international aid and the wages sent home by Basotho miners working in South Africa.

The largest mass market for cannabis products in the region is undoubtedly South Africa. It seems that there exists a kind of South African "cannabis complex" whereby some areas have specialized in producing cannabis in order to supply the consumer markets, most notably those in the large urban areas of Johannesburg (and Gauteng province in general), Durban, KwaZulu-Natal, and Cape Town, Western

Cape province. Although there is little doubt that cannabis is grown throughout South African territory, OGD has identified 4 distinct areas that seem to have specialized in cannabis production as a significant source of income. These are parts of the South African provinces of KwaZulu-Natal and Eastern Cape (the former Transkei), as well as the two small independent states of Swaziland and Lesotho, which are in reality highly dependent on South Africa, both politically and economically. It must be noted that the increasing specialization of these countries' agricultural sector in cannabis production for the South African market reinforces their dependency vis-a-vis their powerful neighbor (see below).

The information on cannabis cultivation presented in this article is based on an OGD field study and on various reports produced by the Lesotho Highlands Development Authority (LHDA) prior to the construction of the large Mohale hydroelectric dam (in Maseru and Thaba-Tseka districts), which will produce electricity for Lesotho and provide water to six South African provinces, including Gauteng Province (Johannesburg). OGD has interviewed six cannabis growers (hereafter identified as "OGD-growers") whose lands in the eastern, mountainous region of Maseru district will be flooded by the Mohale dam. Yet the resulting data is not entirely satisfactory, since it was obtained from a limited sample of growers living solely in the zones affected by the construction of the dam. Strictly speaking, therefore, it cannot be viewed as reflecting the situation prevailing at national level. In addition, although cannabis cultivation is widespread in the mountains, and although all residents of the zones in question (and the country at large) are aware of this fact (as are all local, national and international authorities) - in short, although cannabis production is an open secret and enjoys de facto decriminalization - it nevertheless remains a very private activity. The

LHDA points out that growers are very reticent to discuss the issue, and that it could not gain official access to their fields in order to establish its estimates (in spite of the fact that the LHDA is seriously considering including cannabis revenues in the compensation plan for residents of flooded zones). This information nevertheless provides a fairly detailed picture of the situation of cannabis crops in some farms of the country's mountain regions that furnish, according to all sources, the vast majority of the national harvest.

A Historical Tradition

The first historical record of cannabis in what is now Lesotho dates back to the 16th century. According to historian Stephen Gill, oral tradition has handed down the story of a "colonizing" use of marijuana by the Koena people. The Koena group moved from the northeast of what is now Mpumalanga province (the former Orange Free State) and settled in Lesotho around 1550 (thereby becoming one of the ethnic components of the Basotho group today) by "purchasing" land from San tribes (the earliest inhabitants of South Africa, better known today as "Bushmen") in exchange for marijuana. It is nevertheless very likely that the San knew and used cannabis long before the Koena arrived, these latter simply providing it in great quantity. Furthermore, Gill notes that in the nineteenth century - shortly after the bases of the Kingdom of Lesotho were firmly established by King Moshoeshoe I and the local populations began to depend more on agriculture than on livestock - marijuana figured among the main staples grown in Lesotho, along with sorghum, gourds, and beans.

This historical background suggests why matekoane is now one of the seven plants most often cited by mountain dwellers for their curative and magic qualities. Rural people still use marijuana to treat ailments like heartburn, high blood pressure, and "nerves". It is also used to rid horses and donkeys of parasitic worms (papisi in Sesotho). Two of the six OGD-growers also claimed to smoke marijuana in order to "get strength" and work harder, one of them saying that it stimulated his appetite. According to other sources questioned by the OGD (a psychiatrist and members of a prevention/rehabilitation NGO), these two "utilitarian", or functional, properties are ascribed to matekoane by a high proportion of users throughout Lesotho, both urban and rural.

Among the most "traditional" segments of Basotho society today (i.e., mountain dwellers), marijuana is a medicine considered to have various virtues. But alongside this medicinal status, the field study showed that the general public partly uses the plant for utilitarian or recreational ends not recognized by local traditional medicine.

Nowadays, cannabis is grown almost everywhere in the country, even on small plots in the capital, Maseru. However, the main growing regions are found in the high mountain zones in the center and east of the country, as well as in the western foothill region. Plantations are generally situated in the valleys, where numerous streams and rivers drain the mountains.

Marijuana is Essential to Survival in the Mountains

According to all sources interviewed during the field study, cannabis production is most prevalent in the following districts:- Berea: production occurs in the foothills and mountains located in the east of this district:- Mokhotlong: the eastern sector of this mountainous district (a zone stretching east and south from the Moremoholo River Valley, and including the district capital, Mokhotlong) is part of a region known for its high-quality marijuana ("first grade").

This region also covers parts of Thaba-Tseka and Qacha's Neck districts (see below). The top-grade marijuana is shipped to Durban in South Africa, where it is probably marketed and exported under the name "Durban Poison" (notably to the Netherlands). The western sector of Mokhotlong district yields marijuana of lesser quality;- Thaba-Tseka: whereas the mountainous western sector produces "second-" and "third-grade" cannabis, the equally mountainous south and east belong to the "first grade" production zone mentioned above;- Qacha's Neck: This basically mountainous district belongs almost entirely to - indeed, is the heart of - the 1st Grade cannabis region. The mountains to the west, however, apparently produce 2nd and 3rd grade quality.

The spread and almost universal presence of cannabis crops in every small mountain farm - mountains occupying the largest part of Lesotho, considered as be the only country in the world with an altitude that never drops below 1,000 meters - is also due to soil degradation. Rural dwellers represented 80% of the nation's estimated 2.1 million inhabitants in 1995, but in that same year agriculture supplied less than 15% of Lesotho's GDP, as compared to 25% four years earlier. The beauty of Lesotho's mountains should not mask the serious soil erosion. According to Gill, this erosion accelerated in the early nineteenth century, when areas devoted to grain crops were significantly increased, notably in the lower fields, in order to profit from attractive prices on the international market. These fields were left fallow less and less often, becoming poorer and poorer, while livestock was sent to higher pastures. Every year torrential rains have therefore washed away a little more topsoil from mountains no longer protected by bush (cut for firewood) or grass (overgrazed by the ever-increasing livestock).

Country dwellers consider cattle to be a very important cultural and economic resource, to the extent that all government programs designed to limit its growth and halt overgrazing have failed. The country's population, meanwhile, has grown steadily since the early twentieth century (demographic growth was estimated to be 2.6% per year in 1993). The upshot is that today only 9% of the total surface area of the country is arable land, and it is estimated that an additional 1,000 hectares become inadequate for cultivation each year due to erosion. At this rate, only 8% of Lesotho will be arable in 2001. Note that the sources consulted use "arable land" to refer to fields of marketable crops like grains or beans. Cannabis, meanwhile, can grow in highly depleted soil. The people's two main reactions, historically, to insufficient land have been emigration to South Africa (starting in 1900) and cultivation of cannabis as an export crop (which probably spread sometime later). These two sources of revenue now drive the rural economy.

Emigration has had a distinct if hard-to-quantify impact on the drug situation in Lesotho. Money sent home by emigrant relatives represents the number two source of income for mountain-dwelling households by supplying, according to the LHDA, 38% of the total. In 1993, authorities in Pretoria estimated that 89,400 Basothos worked in South African mines, whereas in 1991 the Maseru authorities estimated that 126,000 Lesotho nationals lived abroad, South African mines, although still the main employer of Basothos, have conducted numerous layoffs in recent years, and are continuing to reduce personnel. Laid-off Basothos do not all return home, but it is probable that some have done so, adding more mouths to feed from over cultivated land, which most likely spurs the (not quantifiable) extension of cannabis crops. Another potential measure -perhaps in addition to extending cultivation - in order to face the new situation characterized by fewer remittances and more people to feed, would be to add value to crops. This trend was noted by a source who declared that he observed that more and more cannabis growers were packaging their produce themselves in the form of ready-to-smoke cigarettes prior to selling to dealers.

According to Gill, the commercial cultivation of cannabis in Lesotho increased considerably from the mid-1980s onward, The LHDA's estimates suggest that households in the Mohale dam zone currently draw 39% of their annual income from agricultural activities. Nearly 50% of that agricultural income (personal consumption included) comes from the sale of cannabis. Cannabis is cultivated in the same way as other crops. Farming in Lesotho's mountains is not modern but is based on rainfall And except for matekoane, crops are mainly destined for personal consumption. Mountain farmers use very little fertilizer (not even natural, like the manure that exists in abundance), pesticides or fungicides, all products of which they remain wary (only 8% of farmers questioned by the LHDA use them).

Mountain agriculture, and cannabis crops in particular, seems to obey the following model: little investment, little risk, low returns. This model appears adapted to the poor mountain soil, which, even with more intensive input, would not yield returns justifying the needed investment. That, at least, is the opinion of local farmers as reported by the LHDA, which does not entirely agree with them.

Whatever the case, cannabis is an indispensable part of the precarious but real equilibrium maintained by mountain farms. Studies by the LHDA, based on low estimates. show that the extremely high value of matekoane means that it supplies nearly half of all agricultural income even though it covers only 10%

of land under cultivation, The LHDA estimates the profit from a hectare of corn to be 209 malotis (M209), as compared to M354 for a hectare of wheat, M493 for a hectare of peas and M4,379 for a hectare of marijuana. It is thus probable that most mountain farms in Lesotho grow a "cluster" of crops, the majority of which are for personal consumption, the sole cash crop being matekoane.

According to available information, all of the cannabis grown in Lesotho comes from small peasant farms in the regions listed above. Various sources indicate that cannabis is usually grown in conjunction with sweet corn, which is the staple crop of Basotho peasants, as well as the basis of their diet. Some cannabis is nevertheless grown as a single crop in more isolated regions, on surface areas that might be as large as five hectares, according to OGD-growers. When planted as a single crop, the size of the OGD-growers' cannabis field is never less than three hectares, which is also the average size of their corn fields. It is worth noting that other sources, generally well-informed on rural life, claim that single-crop cannabis fields are only very rarely larger than one hectare, It is possible that OGD growers have exaggerated the size of their fields thinking (wrongly) that they would obtain more compensation money from the LHDA. According to the studies conducted by the LHDA, the vast majority of mountain farmers work their own land. Some sharecropping and tenant farming exist, but remains marginal. The conclusion is that cannabis production is mainly an economic activity of small owner-farmers.

Planters sow cannabis between mid August and early October, that is to say during the southern spring. Harvesting occurs at the end of the summer, between February and April. Most of the harvest is sold during winter, generally in July. Given the important and increasing supply, winter prices offered by dealers are low (M200 to M300 per bag). Much better prices can be negotiated between November and January (M400 to M600) when the previous year's marijuana stocks are low. Thus, farmers who are able to stock part of their harvest can increase profits by selling during the months when prices are highest. Cannabis therefore constitutes a form of savings for Basotho producers.

Cannabis is sown with seeds obtained from the previous harvest or bought from a neighbor. In both mixed and single-crop fields, matekoane is sown directly in the field where it matures (reportedly, nurseries and transplanting are not employed, as they often are in West Africa). Care involves weeding the plot and, in a few cases, applying manure and irrigating. Women generally perform these tasks, but there are many phases that involve all members of the family, as is always the case at harvest time, when men, women, and children work together. Harvesting and packing (see below) are sometimes the occasion for "work parties" where neighbors and paid workers join in, although this system would not seem to be the rule.

The first harvest, probably carried out in January, is done on what farmers call majaja. According to accounts provided by the OGD-growers, majaja comes from the same seeds as "the real matekoane", yet bears no flowers or seeds. It can then be deducted that majaja is the male plant of cannabis. The majaja harvest therefore represents a thinning of the plots, leaving only the female plants. Whereas in other countries such as Morocco this thinning is normally viewed as a task designed to improve the final product, it seems that in Lesotho it has a commercial goal, namely to market another full-fledged product. It was difficult to obtain information on majaja, which growers distinguish from matekoane in terms of labor (only the leaves of majaja are retained) and income (majaja earns less). The leaves of male plants are separated from the stalks and sold in bags. It is probable that Sesotho majaja is the substance sold in South Africa under the name of maajut, poor quality marijuana basically used for smoking with Mandrax in what is called "white pipe".

The main harvest of "real matekoane" (which contains seeds and flowers) begins in February and may continue until April; depending on weather conditions and geographical situation. The harvested plants are carried to the farmhouse where they are generally left to dry outside, on the ground. The flowers are then separated from the stalks. The flowers are stuffed into bags (probably together with a certain amount of leaves) which normally contain 50 kilograms of corn and which constitute the unit of sale in the fields.

A Basotho source whose work entails frequent contact with the mountain-dwelling communities stated that in recent years increasing numbers of growers in the Qabane river valley (on the eastern edge of Mohale's Hoek district) were rolling their matekoane into cigarettes prior to selling it, thereby adding value that increased prices. According to this source, the task is carried out by women, and involves no machinery. If this innovation extends to other areas of the country (it was not mentioned by either the LHDA studies or the OGD-growers), that would represent another sign of the already obvious decriminalization of the cultivation and, to a lesser extent, the sale of cannabis in Lesotho.

Above all, however, it might indicate a growing specialization in cannabis crops in certain areas, with a concomitant monetization of the economy, insofar as packing even a part of the marijuana harvest in the form of cigarettes probably requires a great deal of time. That time would no longer be available for other tasks generally allotted to women, for example cultivating food crops, especially vegetables. An hypothesis may be made that if these tasks are abandoned in favor of rolling marijuana into cigarettes, rural households will increasingly depend on commercial networks rather than their own labor for food.

According to the OGD-growers, who live relatively far from the country's borders, the harvest is usually taken from the production zone by traffickers who employ automobiles (usually 4-wheel drive vehicles, known as "bakkies" in Lesotho and South Africa). The harvest for a given zone is first brought to a spot accessible by car, at the buyer's expense According to the OFD-growers, the purchasers are sometimes Basotho but usually Zulu or Xhosa (two South African ethnic groups) and pay mountain dwellers (usually women) to transport the matekoane harvest to the assembly point. Purchasers sometimes also rent the growers' donkeys to get the harvest to more distant assembly points. In other regions of Lesotho, for example in Mokhotlong, Thaba-Tseka, and Qacha's Neck districts (in Eastern Lesotho, near the border with South Africa's KwaZulu-Natal province) caravans of donkeys and "porters" carry the marijuana across the border, probably into Zulu villages. From there it is shipped on to Durban, usually in collective taxis.

It should be noted that the temporary hire of farmers as porters, and the rental of their donkeys, are advantageous arrangements for growers, because it means that transporting the cannabis harvest provides another distinct source of income in addition to straightforward cultivation. It proved impossible, however, to obtain details - even approximate - on the scope of the income thus generated.

Trafficking and its Effects

The tense political atmosphere reigning in Lesotho in the southern winter of 1997 made the study into trafficking, networks rather difficult. No one wanted to risk even hinting on the possibility that local prominent citizens might be implicated in trafficking in any way whatsoever. It would nevertheless be quite surprising, at least as far as cannabis trafficking goes (which entails little social or penal condemnation), if traffickers enjoyed no bureaucratic or military - or, indeed, political - protection. Non-Basotho sources moreover declared that some Basotho politicians more or less openly viewed cannabis revenues as an unofficial but useful boost to the country's balance of payments. A Basotho civil servant, when confronted with the contradictions in his comments, finally admitted that, given the political situation, "civil servants don't dare take action because they can't foresee the consequences of their acts". The September 1998 insurrection in Maseru, followed by a joint military intervention by South Africa and Botswana, resulted in the destruction of the capital. This further weakened the capabilities of Basotho law enforcement and severely restricted the country's independence.

Like their counterparts in other Southern African countries, Basotho officials often place all the blame on "foreigners", who are convenient scapegoats because they are politically neutral. Basothos are even reticent to offer detailed information on compatriots arrested in Lesotho itself on drug charges. South Africans, on the other hand, are accused of fomenting cannabis production in the mountains, while Nigerians are blamed for the growing (if still limited) use not only of cocaine and but also synthetic drugs like LSD and ecstasy (in which Nigerian involvement is improbable). The Indo-Pakistani community, meanwhile, is suspected of extensively trafficking Mandrax, although no scandal has ever come to light (at least publicly) to confirm such suspicions. Although some of these accusations may not be totally unfounded, they help disguise local responsibility for the demobilization and disorganization of drug enforcement measures, not to mention the protection and perhaps even collusion required for certain operations.

All the cannabis grown in landlocked Lesotho is exported to South Africa, at least initially. There are two main export routes. One heads west and north toward Bloemfontein and Ficksburg, then on to Johannesburg. This is the route taken by 2nd and 3rd grade matekoane grown in western and central Lesotho. Transportation is usually done by motor vehicles (cars and trucks). It is likely that at least part of the marijuana is centralized in the towns of Maseru and Mafeteng prior to being shipped across the border. It is also probable that these towns have relatively large storage facilities.

The other route leads to Durban, the destination for 1st grade marijuana grown in the eastern districts of Mokhotlong, Thaba-Tsella, and Qacha's Neck (see above). According to OGD's sources, high-grade matekoane often arrives in KwaZulu-Natal villages on the backs of donkeys and porters. It is likely that cross-country motor vehicles are also used. Once in South Africa, Basotho marijuana is taken to Durban townships by collective taxis (many of the taxi firms in townships around Johannesburg, Durban and Cape Town are owned by dealers in dagga - as marijuana is usually called in South Africa). Once in Durban, the cannabis will be packaged and sold on the national market or exported to Europe (until now it seems mainly to the Netherlands and the United Kingdom) or even to North America, often mixed with marijuana grown in KwaZulu-Natal.

According to the available information, these two routes are mainly used by networks of South African traffickers, who supply their country's urban markets. Yet there also exist parallel marijuana networks supplying Basotho miners working in South Africa. Most miners in South Africa are known to make "utilitarian" use of marijuana, and sometimes Mandrax, to crank themselves up for work and to "chill out" afterward. The South African police have raided hostels where Basotho miners stay and has found sacks of marijuana. According to South African and Basotho police officials, Lesotho marijuana is highly appreciated by users all over South Africa.

It is worth noting that the isolation of the central and eastern mountain regions of Lesotho makes aircraft the best titans of transportation. Some thirty small airfields are scattered across the country. It seems likely that certain airfields are used to ship middling-size quantities of marijuana to Masern or other urban centers, even though most sources questioned in Lesotho, including the police, remain sceptical. Although no concrete evidence has ever come to light, it would hardly be surprising if small aircraft flew marijuana directly into South Africa.

It seems that a mutually fuelling relationship exists between the cannabis trade and other kinds of illicit activity in Lesotho:

The first activity concerns stolen cars (like everywhere else in Southern Africa). Cars stolen in South Africa and beyond are sold cheaply in Lesotho. Vehicles are also stolen in Lesotho for re-sale abroad (primarily South Africa and Zambia). Once construction began on the Mohale dam, South African expatriates working on the site in the mountain zone were often victims of car theft, sometimes also losing their lives. Ever since, many South Africans working in Lesotho carry weapons, and car thefts have become the main concern of South Africa's High Commission in Maseru. Many members of the Chinese community (which control the small garment industry) also carry guns. The Chinese, known as very tough bosses, are detested by the locals and have been the victims of violent attacks and car theft (anti-Chinese riots took place in January 1998).

The second smuggling activity concerns stolen livestock (cows, sheep, and goats stolen in Lesotho for resale in South Africa, and vice versa). As already noted, livestock is a sign of wealth among Basothos (many of whom also live on the other side of the border), so there are cows everywhere. But there is also a constant desire to own more. Farmers are arming themselves defensively against thieves. Furthermore, even though marijuana trafficking is generally non-violent, the police claim that some producers have armed themselves against enforcement agents. Moreover, in the spring of 1997, marijuana smugglers in a national park on the country's northern border attacked South African hikers.

It can be deducted from the above that the cannabis trade is partly based on barter arrangements and that it is linked, at least indirectly, to the proliferation of small arms in Lesotho (and, as an OGD study has shown, throughout Southern Africa).

"Laundering" Cannabis Revenue

Cannabis cultivation and trafficking probably constitute two of rural Lesotho's more widespread and rewarding economic activities. Growers use marijuana income for everyday expenditures, notably for sending their children to school (secondary education is expensive in Lesotho). It is hard to speak of money laundering in this instance, since income from matekoane is an integral part of mountain farmers' economy. Moreover, South African and Basotho traffickers go to the mountains and buy directly from the growers, which means that the revenues generated by cannabis in the countryside are broadly distributed, rather than concentrated in a few wholesalers' hands as is the case elsewhere, for instance in near-by Swaziland. Concentration occurs among South African traffickers and probably also among Basothos in the urban zones in western Lesotho, although no trustworthy information is available on this latter group.

An unusual form of "laundering" will certainly take place in the context of compensation for lands flooded by the Mohale dam. Sources claim that the LHDA is working on a project in association with many foreign institutional investors to take into account income generated by matekoane when it comes to compensate for losses incurred by flooding farmlands. Therefore, top level institutions judge -

correctly, we feel - illicit crops to be a key part of economic life in Lesotho's rural heartland.

SOUTH AFRICA

"When you look at the statistics, you have to ask what the police are doing. Seizures are completely out of proportion with the explosion in the country's drug market" This disenchanted commentary from a high-ranking European civil servant posted to South Africa illustrates the widespread feeling that the authorities remain overwhelmed by the drug phenomenon in the South Africa, despite the harder line adopted recently by the government following the June 1999 election of Thabo Mbeki to succeed Nelson Mandela as president. During the first quarter of 1999, for example, less than 90 kilograms of cocaine hydrochloride were seized in South Africa, compared to over 170 kg during the same period the previous year (more than 620 kg of cocaine hydrochloride were seized for the whole of 1998). In addition, only 654 kg of cannabis were confiscated in the province of Gauteng (Johannesburg/Pretoria) during the first quarter of 1999, compared to 12.3 metric tons during all of 1998. That is in a country which remains one of the world's principal cannabis producers (various sources estimate the surface area under cannabis cultivation at between 35,000 and 80,000 hectares).

ALGERIA

...In 1999, according to official statistics, 2.5 metric tons of hashish and 5 kilograms of cocaine were seized. Even if the quantities declared represent only a fraction of what is annually seized they are still five times higher than in 1997 and three times higher than 1998. A total of 5,300 people, including 1,000 foreigners, were arrested for drug trafficking. Drug use, encouraged by political violence and the decline of authority, in addition to poverty and social exclusion, affects an increasing number of young people depressed by the lack of opportunities or trying to combat stress and fear by seeking solace in artificial happiness. Supplies to neighborhoods that have been abandoned by the state either come from outside the country - across a hard-to-monitor land border 6,700 kilometers long, or via networks based in Europe - or are home-produced: cannabis crops have been destroyed in both the north and south of the country, and although the amounts are still small, they show that local supply networks are in the process of being set up.

Shared Trafficking

In order to understand the drug situation at the start of 2000, we must look back to the setting up of the networks when the conflict was at its height. Although the military obtained their drugs on the legitimate market, the fundamentalists and some unemployed people entered the black markets, which are supplied from Europe or Morocco or even, in the case of cannabis derivatives, from local crops: "we often find cannabis grown in backyards or on balconies", an officer from the drug squad said at the start of 1999. The squad is only a shadow of its former self since most of its staff had been reassigned to more urgent security jobs.

Sparsely planted cannabis crops can be seen in the southern oases and in the dense forests covering the Medea mountains in the south-west and the Tlemcen mountains in the west, which were for a long time rendered inaccessible to the military by rebels from the Armed Islamic Groups (GIA). Morocco is still the main source of supply. In fact, in the early 1990s, Algeria emerged as an alternative route for exporting Moroccan hashish to Europe. The closing of the border between the two countries in 1994 hit those networks for a time. However, there were no attacks in the border region between 1994 and 1998, so the networks have been flourishing again.

In 1999, the hashish route varied very little. Beginning at the Moroccan border, in a no man's land marked out by the armies of both countries (in either side, it snakes across plains and plateaus to Annaba in the far east of Algeria, passing through Oran, Algiers and Constantine on the way. From its "home" in the Rif mountains of northern Morocco, the hashish passes through the roadblocks of the police and army as well as the "false roadblocks" (faux barrages) of the CIA. Naturally, the traffickers have to pay off both sides in the conflict that has ravaged Algeria for the past ten years.

Some of the drugs are sold on local markets, while the rest continue on to Europe, via Tunisia and Italy, or directly to Marseilles. According to one attentive observer, since the Algerian civil war was stepped up, it is easier to go through northern Algeria, taking advantage of existing networks, the disarray of the government, and the armed gangs' excellent knowledge of the terrain, than to take the traditional desert routes across West Sahara or Libya. But southern Algeria, away from the scene of fighting, was still being used as a "caravan route" for ferrying drugs such as hashish and heroin between western Africa and Sudan, The few seizures reported by the authorities confirm this evaluation of the situation.... Although hashish reigns supreme, psychotropic drugs are gaining ground, making up about 20% of the drugs the youngsters acknowledge using.

NIGERIA

...The fight against cannabis crops - Nigeria is one of the largest producers of marijuana in Africa, with 80 metric tons seized between 1994 and 1999 - is the only field where the NDLEA obtained a modicum of success during the last two years. Thus in late August 1998, hundreds of hectares of cannabis were eradicated in a national park in Ondo State, in the south-west of the country. During the previous three months, 80 small plantations were destroyed.

It seems that traffickers in hard drugs have been much less targeted by Nigerian law enforcement, while a number of cases abroad have implicated Nigerian traffickers.

CONGO

...Meanwhile, the profits generated by cannabis trafficking are one of the reasons for the persistence of violent clashes in Congo's rural areas, especially in the Pool region in the south-west of the country.

Marijuana and cocaine networks

Colonel Nkou, Police Chief for, the Poto-Poto neighborhood, confirmed to the OGD correspondent that his men continuously arrested traffickers in possession of several kilograms, sometimes several dozen kilos, of marijuana. Supply is so high that a small 10-gram ball of marijuana sells for only 100 CFA francs (about 20 US cents). Nkou also revealed the launching of a vast operation in July to put an end to drug dealing in the capital. The operation reportedly consists of eradicating all marijuana crops located in the surrounding villages. But information obtained by the OGD correspondent suggests that this has not been the case. One source declared: "The police are poorly equipped. But more importantly they are afraid, because it is well known that it is some of the clan militiamen defeated in the war who have withdrawn to the village to grow cannabis alongside their parents". However, the civil wars in both Brazzaville and neighboring Kinshasa have also facilitated the development of the trade in hard drugs, especially cocaine....

Rural warfare for cannabis

After experiencing the various stages that drug plants have historically experienced in other world regions

- a means to survive for poor farmers and a source of funds for militias in urban and rural areas) -, cannabis has recently assumed a truly geopolitical dimension in the Congo by entering a new phase: becoming the stakes of a conflict between ethnic groups and political factions for the control of production. In the south of the country, notably in the towns of Mayama, Kindamba and Mindouli, Pool region, which border on the Bouenza region - an electoral stronghold of the deposed president Lissouba and of his ephemeral prime minister Bernard Kolelas, the former mayor of Brazzaville - the Angolan soldiers, locally known as "amigos", in charge of the region's security on behalf of the Sassou government, have a marked taste for cannabis and alcohol. They barter both drugs for their food rations with the local people. The Angolans get along well with the former Ninja and Cokoye militiamen supporting Lissouba who have returned to their home region, but they are disliked by the Cobra militiamen of President Sasson and by the Congolese police with whom they sometimes have violent arguments.

But during the last quarter of 1998 the quarrels have turned into rural warfare, Indeed, in the Pool region, south west of Brazzaville, the capital, violent clashes have set the "Ninja" militias of Bernard Kolelas against the military and the former "Cobra" militias of current president Denis Sasson Nguesso. It all started with the summary execution, ordered by police captain Kamar, a former collaborator of Kolelas who rallied the Sassou Nguesso government, of three ninja leaders in the community of Mindouli, 180 kilometers south west of Brazzaville in the Pool region. The bodies were buried standing by the highway with their heads emerging from the ground as a warning for other ninja fighters. Following the victory of the Democratic and Patriotic Forces (FDP) of Sassou, many ninjas fled to Mindouli, where, posing as farmers, they started growing cannabis on a large scale. Mindouli became the largest cannabis production area in the zone running along the Congo-Ocean railway, and the Pool therefore turned into the Congo's number one marijuana producer ahead of the Plateaux and Cuvette regions. Captain Kamar, a prominent marijuana producer and trafficker under the Lissouba government, has carried through his activities under the new regime with the complicity of senior civil servants and help from some former ninjas. But because he had failed to pay for the cannabis some ninjas had delivered, the latter started harvesting the land of the "boss" for their own profit. This is what triggered the exemplary execution of the three ninjas referred to above. Other ninjas were not impressed and took up their arms in order to retaliate. On August 29, 1998, they attacked the Mindouli police station, killing the police commissioner and regional prefect, whom they alleged was a member of the Kamar network. A journalist working for a radio close to the government, Michel Mampoya, who happened to be in Mindouli, was also killed. President Sassou made matters worse by sending in the military and Cobra militiamen, which started full-blown rural guerrilla warfare. In a few days, the ninjas became the masters of the region after inflicting severe losses on government forces. In the village of Goma Tse-Tse in September, they murdered the prefect and his wife. In Kibossi, they killed the chief of the village whose only crime was to have advised the minister of culture and sport, Ernest Ndala, to leave the area before the arrival of the ninjas. The villagers, who supply the rebels with food and fighters, have become a target for the military and the Cobras, and rapes, pillaging and stealing of cannabis harvests are now commonplace. The train, which was confiscated by the government in order to transport troops to the Pool, came back to Brazzaville loaded with the loot, including sacks of marijuana. The drugs are immediately disposed of at very low prices either at the train station itself or in working class neighborhoods. Although the military has gained control of several villages in the Pool, the rebels are far from being subdued. On November 14, 1998, six priests, who were members of the Mediation Committee set up to resolve the conflict, were murdered in Mindouli. The government blamed the murder on the ninjas, but observers were not convinced. In mid-December, the fighting left 60 people dead. The passenger train, which normally supplies the capital with agricultural produce, no longer stops at stations in the Pool. As a result, marijuana prices have shot up in Brazzaville and the weight of joints has gone down. The opposition in exile and the government blame each other for a terrible situation that has older causes. The Lari, the majority ethnic group in the country to which former prime minister Kolelas belongs, are traditional cannabis producers who are very much attached to their land. For 30 years they have been accused of being "against the revolution" because they rejected the marxism of the

Northerners now back in power. It is likely that the intervention of the military in the region, where many people have fled their villages to take refuge in the forest, will result in never-ending guerrilla warfare with the civil population governed by the ninjas and funded by expanding cannabis crops. The government, not surprisingly, accuses Kolelas of controlling the trade in marijuana in order to buy arms for his supporters. But in Brazzaville, there is information about the very lucrative relations maintained between some dignitaries of the Sassou government and Lebanese businessmen who import dozens of kilograms of cocaine into the Congo.

MAURITIUS

...Rich Man's Marijuana and Poor Man's Heroin

The social and ethnic structure of the island results in varying and complex consumption patterns. Cannabis is regarded as the drug of the wealthy, in the same way as cocaine. Its price varies, but it is relatively expensive at about $10 a bag for local people, or $5 a joint when sold to tourists - often by "rastas" who haunt the big tourist centers, casinos and beaches. Although ganja consumption is overshadowed by the alarming statistics on heroin use, that does not mean it is insignificant: it takes place on a large scale, but mainly involves "protected population groups".

In this respect, it is worrisome to note that it is increasingly the better-off who use the drug, and that young people and the unemployed can no longer afford it. Mainly for financial reasons, they are more likely to use brown sugar, medical drugs, glue or solvents....The outcome of government policy, which is aimed at small-scale dealers and users, is to congest Mauritian jails: people convicted of drug offences make up 70% of the prison population.

HEALTH BENEFITS OF CANNABIS IN AFRICA

From promoting appetite in chemo patients to potentially protecting the brain from trauma caused by a concussion, there are plenty of medical marijuana uses.

The African public largely supports the legalization of medical marijuana. At least 84% of the public believes the drug should be legal for medical uses, and recreational pot usage is less controversial than ever, with at least 61% of Africans in support.

Even though some medical benefits of smoking pot may be overstated by advocates of marijuana legalization, recent research has demonstrated that there are legitimate medical uses for marijuana and strong reasons to continue studying the drug's medicinal uses.

Even the NIH's National Institute on Drug Abuse lists medical uses for cannabis.

There are at least two active chemicals in marijuana that researchers think have medicinal applications. Those are cannabidiol (CBD) — which seems to impact the brain without a high— and tetrahydrocannabinol (THC) — which has pain relieving properties and is largely responsible for the high.

But scientists say that limitations on marijuana research mean we still have big questions about its medicinal properties. In addition to CBD and THC, there are another 400 or so chemical compounds, more than 60 of which are cannabinoids. Many of these could have medical uses. But without more research, we won't know how to best make use of those compounds.

More research would also shed light on the risks of marijuana. Even if there are legitimate uses for medicinal marijuana, that doesn't mean all use is harmless. Some research indicates that chronic, heavy users may have impaired memory, learning, and processing speed, especially if they started regularly using marijuana before age 16 or 17.

For some of the following medical benefits, there's good evidence. For others, there's reason to continue conducting research.

A recent report by the National Academies of Sciences, Engineering, and Medicine said there was definitive evidence that cannabis or cannabinoids (which are found in the marijuana plant) can be an effective treatment for chronic pain.

The report said that is "by far the most common" reason people request medical marijuana.

That same report said there's equally strong evidence marijuana can help with muscle spasms related to multiple sclerosis.

Other types of muscle spasms respond to marijuana as well. People use medical marijuana to treat diaphragm spasms that are untreatable by other, prescribed medications.

It doesn't seem to harm lung capacity, and may even improve it.

There's a fair amount of evidence that marijuana does no harm to the lungs, unless you also smoke tobacco. One study published in Journal of the American Medical Association found that not only does marijuana not impair lung function, it may even increase lung capacity.

Researchers looking for risk factors of heart disease tested the lung function of 5,115 young adults over the course of 20 years. Tobacco smokers lost lung function over time, but pot users actually showed an increase in lung capacity.

It's possible that the increased lung capacity may be due to taking a deep breaths while inhaling the drug and not from a therapeutic chemical in the drug.

The smokers in that study only toked up a few times a month, but a more recent survey of people who smoked pot daily for up to 20 years found no evidence that smoking pot harmed their lungs, either.

The National Academies report said there are good studies showing marijuana users are not more likely to have cancers associated with smoking.

It may be of some use in treating glaucoma, or it may be possible to derive a drug from marijuana for this use.

One of the most common reasons that states allow medical marijuana use is to treat and prevent the eye disease glaucoma, which increases pressure in the eyeball, damaging the optic nerve and causing loss of vision.

Marijuana decreases the pressure inside the eye, according to the National Eye Institute: "Studies in the early 1970s showed that marijuana, when smoked, lowered intraocular pressure (IOP) in people with normal pressure and those with glaucoma."

For now, the medical consensus is that marijuana only lowers IOP for a few hours, meaning there's not good evidence for it as a long term treatment right now. Researchers hope that perhaps a marijuana-based compound could be developed that lasts longer.

It may help control epileptic seizures.

Some studies have shown that cannabidiol (CBD),

another major marijuana compound, seems to help people with treatment-resistant epilepsy.

A number of individuals have reported that marijuana is the only thing that helps control their or their children's seizures.

However, there haven't been many gold-standard, double-blind studies on the topic, so researchers say more data is needed before we know how effective marijuana is.

It also decreases the symptoms of a severe seizure disorder known as Dravet's Syndrome.

During the research for his documentary "Weed," Sanjay Gupta interviewed the Figi family, who treated their 5-year-old daughter using a medical marijuana strain high in cannabidiol and low in THC.

The Figi family's daughter, Charlotte, has Dravet Syndrome, which causes seizures and severe developmental delays.

According to the film, the drug decreased her seizures from 300 a week to just one every seven days. Forty other children in the state were using the same strain of marijuana to treat their seizures when the film was made — and it seemed to be working.

The doctors who recommended this treatment said the cannabidiol in the plant interacts with brain cells to quiet the excessive activity in the brain that causes these seizures.

Gupta notes, however, that a Florida hospital that specializes in the disorder, the American Academy of Pediatrics, and the Drug Enforcement agency don't endorse marijuana as a treatment for Dravet or other seizure disorders.

A chemical found in marijuana stops cancer from spreading, at least in cell cultures.

CBD may help prevent cancer from spreading, researchers at California Pacific Medical Center in San Francisco reported in 2007.

Other very preliminary studies on aggressive brain tumors in mice or cell cultures have shown that THC and CBD can slow or shrink tumors at the right dose, which is a strong reason to do more research.

One 2014 study found that marijuana can significantly slow the growth of the type of brain tumor associated with 80% of malignant brain cancer in people.

Still, these findings in cell cultures and animals don't necessarily mean the effect will translate to people — far more investigation is needed.

It may decrease anxiety in low doses.

Researchers know that many cannabis users consume marijuana to relax, but also that many people say smoking too much can cause anxiety. So scientists conducted a study to find the "Goldilocks" zone: the right amount of marijuana to calm people.

According to Emma Childs, an associate professor of psychiatry at the University of Illinois at Chicago and an author of the study, "we found that THC at low doses reduced stress, while higher doses had the opposite effect."

A few puffs was enough to help study participants relax, but a few puffs more started to amp up anxiety. However, people may react differently in different situations.

THC may slow the progression of Alzheimer's disease.

Marijuana may be able to slow the progression of Alzheimer's disease, a study led by Kim Janda of the Scripps Research Institute suggests.

The 2006 study, published in the journal Molecular Pharmaceutics, found that THC (the active chemical in marijuana) slows the formation of amyloid plaques by blocking the enzyme in the brain that makes them. These plaques kill brain cells and are associated with Alzheimer's.

A synthetic mixture of CBD and THC seems to preserve memory in a mouse model of Alzheimer's disease. Another study suggested that a THC-based prescription drug called dronabinol was able to reduce behavioral disturbances in dementia patients.

All these studies are in very early stages, though, so more research is needed.

The drug eases the pain of multiple sclerosis.

Marijuana may ease painful symptoms of multiple sclerosis, according to a study published in the Canadian Medical Association Journal.

Jody Corey-Bloom studied 30 multiple sclerosis patients with painful contractions in their muscles. These patients didn't respond to other treatments, but after smoking marijuana for a few days, they reported that they were in less pain.

The THC in marijuana seems to bind to receptors in the nerves and muscles to relieve pain.

It seems to lessen side effects from treating hepatitis C and increase treatment effectiveness.

Treatment for hepatitis C infection is harsh: negative side effects include fatigue, nausea, muscle aches, loss of appetite, and depression. Those side effects can last for months, and lead many people to stop their treatment course early.

But a 2006 study in the European Journal of Gastro-enterology and Hepatology found that 86% of patients using marijuana successfully completed their Hep C therapy. Only 29% of non-smokers completed their treatment, possibly because the marijuana helps lessen the treatment's side effects.

Marijuana also seems to improve the treatment's effectiveness: 54% of hep C patients smoking marijuana got their viral levels low and kept them low, in comparison to only 8% of nonsmokers.

Marijuana may help with inflammatory bowel diseases.

Patients with inflammatory bowel diseases like Crohn's disease and ulcerative colitis could benefit from marijuana use, studies suggest.

University of Nottingham researchers found in 2010 that chemicals in marijuana, including THC and cannabidiol, interact with cells in the body that play an important role in gut function and immune responses. The study was published in the Journal of Pharmacology and Experimental Therapeutics.

The body makes THC-like compounds that increase the permeability of the intestines, allowing bacteria in. But the cannabinoids in marijuana block these compounds, making the intestinal cells bond together tighter and become less permeable.

But the National Academies report said there isn't enough evidence to be sure whether marijuana really helps with these conditions, so more research is needed.

It relieves arthritis discomfort.

Marijuana alleviates pain, reduces inflammation, and promotes sleep, which may help relieve pain and discomfort for people with rheumatoid arthritis, researchers announced in 2011.

Researchers from rheumatology units at several hospitals gave their patients Sativex, a cannabinoid-based pain-relieving medicine. After a two-week period, people on Sativex had a significant reduction in pain and improved sleep quality compared to placebo users.

Other studies have found that plant-derived cannabinoids and inhaled marijuana can decrease arthritis pain, according to the National Academies report.

Marijuana users tend to be less obese and have a better response to eating sugar.

A study published in the American Journal Of Medicine suggested that pot smokers are skinnier than the average person and have healthier metabolism and reaction to sugars, even though they do end up eating more calories.

The study analyzed data from more than 4,500 adult Americans — 579 of whom were current marijuana smokers, meaning they had smoked in the last month. About 2,000 people had used marijuana in the past, while another 2,000 had never used the drug.

The researchers studied how the participants' bodies responded to eating sugars. They measured blood-sugar levels and the hormone insulin after participants hadn't eaten in nine hours, and after they'd eaten sugar.

Not only were pot users thinner, their bodies also had a healthier response to sugar. Of course, the study couldn't determine whether the marijuana users were like this to begin with or if these characteristics were somehow related to their smoking.

While not really a health or medical benefit, marijuana could spur creativity.

Contrary to stoner stereotypes, marijuana usage has actually been shown to have some positive mental effects, particularly in terms of increasing creativity, at least in some contexts. Even though people's short-term memories tend to function worse when they're high, they actually get better at tests requiring them to come up with new ideas.

Researchers have also found that some study participants improve their "verbal fluency," their ability to come up with different words, while using marijuana.

Part of this increased creative ability may come from the release of dopamine in the brain, which lowers inhibitions and allows people to feel more relaxed, giving the brain the ability to perceive things differently.

Cannabis soothes tremors for people with Parkinson's disease.

Research from Israel shows that smoking marijuana significantly reduces pain and tremors and improves sleep for Parkinson's disease patients. Particularly impressive was the improved fine motor skills among patients.

Medical marijuana is legal in Israel for multiple conditions, and a lot of research into the medical uses of cannabis is done there, supported by the Israeli government.

Marijuana may help veterans suffering from PTSD.

In 2014, the Colorado Department of Public Health awarded $2 million to the Multidisciplinary Association for Psychedelic Studies (one of the biggest proponents of marijuana research) to study marijuana's potential for people with post-traumatic stress disorder.

Naturally occurring cannabinoids, similar to THC, help regulate the system that causes fear and anxiety in the body and brain.

Marijuana is approved to treat PTSD in some states already — in New Mexico, PTSD is the number one reason for people to get a license for medical marijuana.

But there are still questions about the safety of using marijuana while suffering from PTSD, which this study — which has taken a while to get off the ground — will hopefully help answer.

Animal studies suggest that marijuana may protect the brain after a stroke.

Research from the University of Nottingham shows that marijuana may help protect the brain from damage from a stroke by reducing the size of the area affected by the stroke — at least in rats, mice, and monkeys.

This isn't the only research that has shown neuroprotective effects of cannabis. Some research shows that the plant may help protect the brain after other types of brain trauma.

Marijuana might even protect the brain from concussions and trauma.

Lester Grinspoon , a professor of psychiatry at Harvard and marijuana advocate, recently wrote an open letter to NFL Commissioner Roger Goodell. In it, he said the NFL should stop testing players for marijuana, and that the league should start funding research into the plant's ability to protect the brain instead.

"Already, many doctors and researchers believe that marijuana has incredibly powerful neuroprotective properties, an understanding based on both laboratory and clinical data," Grinspoon wrote.

Goodell said he'd consider permitting athletes to use marijuana if medical research shows that it's an effective neuroprotective agent.

At least one recent study on the topic found that patients who had used marijuana were less likely to die from traumatic brain injuries.

It can help eliminate nightmares.

This is a complicated one, because it involves effects that can be both positive and negative. Marijuana disturbs sleep cycles by interrupting the later stages of REM sleep. In the long run, this could be a problem for frequent users.

However, for people suffering from serious nightmares, especially those associated with PTSD, this can be helpful, perhaps in the short term. Nightmares and other dreams occur during those same stages of sleep. By interrupting REM sleep, many of those dreams may not occur. Research into using a synthetic cannabinoid — similar to THC but not the same — showed a significant decrease in the number of nightmares in patients with PTSD.

Additionally, even if frequent use can be bad for sleep, marijuana may be a better sleep aid than some other substances that people use. Some of those, including medication and alcohol, may potentially have worse effects on sleep, though more research is needed on the topic.

Cannabis reduces some of the pain and nausea from chemotherapy and stimulates appetite.

One of the most well-known medical uses of marijuana is for people going through chemotherapy. There's good evidence that it's effective for this, according to the National Academies report.

Cancer patients being treated with chemo suffer from painful nausea, vomiting, and loss of appetite. This can cause additional health complications.

Marijuana can help reduce these side effects, alleviating pain, decreasing nausea, and stimulating the appetite. There are also multiple FDA-approved cannabinoid drugs that use THC, the main active chemical in marijuana, for the same purpose.

Marijuana can help people who are trying to cut back on drinking.

Marijuana is safer than alcohol. That's not to say it's risk-free, but cannabis is much less addictive than alcohol and doesn't cause nearly as much physical damage.

Disorders like alcoholism involve disruptions in the endocannabinoid system. Because of that, some people think cannabis might help patients struggling with those disorders.

Research published in the Harm Reduction Journal found that some people use marijuana as a less harmful substitute for alcohol, prescription drugs, and other illegal drugs. Some of the most common reasons patients make that substitution are that marijuana has less negative side effects and is less likely to cause withdrawal problems.

Some people do become psychologically dependent on marijuana, and it is not a cure for substance abuse problems. But from a harm-reduction standpoint, it can help.

Still, it's worth noting that combining marijuana and alcohol can be dangerous, and some researchers are concerned that this scenario is more likely than one in which users substitute a toke for a drink.

Medical marijuana legalization seems to reduce opioid overdose deaths.

While there are a number of factors behind the current opioid epidemic, many experts agree that the use of opioid painkillers to treat chronic pain has played a major role. It's very risky to take powerful drugs that have a high risk of causing overdose and high addiction rates. Marijuana, which can also treat chronic pain, is far less risky.

Several studies have showed that states that allow medical marijuana have fewer opioid deaths. This effect seems to grow over time, with states who pass these laws seeing a "20% lower rate of opioid deaths in the laws' first year, 24% in the third, and 33% in the sixth," according to Stat News.

It's hard to say that deaths went down because of medical marijuana legalization and not other reasons. But because the effect seems to get stronger the longer marijuana remains legal, researchers think marijuana is a likely cause of the decline in opioid deaths.

LESOTHO: AFRICA'S PIONEER WHEN IT COMES TO CANNABIS

For once, the spotlight is not on its big brother, South Africa, but on Lesotho itself: In 2017, this small country legalised medicinal cannabis, making it the first place in Africa to do so! Cannabis News Network took a quick look around this beautiful but poor country and found that not very much had really changed since the legalisation.

Stretching from its capital city of Maseru in the west, and the Drakensberg mountains in the East, Lesotho welcomes visitors with stunning mountain landscapes. As the only country in the world that is entirely above 1,000 m, it is for good reason that Lesotho is known as the kingdom on the roof of Africa.

Alongside the beauty of the landscape, the other thing that leaps out at you is the poverty. Almost half the population has to try to live on just over USD 1.00 per day. Life expectancy is one of the lowest in the world, HIV and Aids remain huge problems.

Growing cannabis in Lesotho

Cannabis has a special status in Lesotho. Since ancient times, it has formed an integral part of the tribal culture, something that its cultural and social acceptance fully reflects. This was the case even before legalisation.

For many farmers in this country, cannabis is the only agricultural product that allows them to earn cash. Many of them were hoping their lives would improve thanks to legalisation, but that never happened.

The legal cultivation of cannabis is limited to two foreign companies. It is a pity and repugnant that the population of Lesotho will not be profiting from the new, legalised market. This trend is unfortunately also showing its face in other parts of the world, as is shown in a three-part report on Cannabis News Network from Colombia.

Chris Ranthini, a young climate activist, sums up the problem in clear terms: "We should not have black laws for some people and white laws for other people. If it is legal to grow cannabis in Lesotho, then let the small farmers grow it."

WAYS LEGAL CANNABIS WILL CHANGE THE WORLD

Marijuana legalization is having profound beneficial effects in the cities in which it is available. We take a closer look at the areas of society that benefit the most.

Marijuana, the budding flower of the cannabis plant, has long held an unfortunate stigma. Once openly cultivated around the world, marijuana fell into its own dark ages, turning consumers and growers in criminals. All too often, marijuana found itself associated with dangerous drugs like cocaine and heroin. As the years passed, awareness of the medical benefits of marijuana has become more widespread sparking debate around the globe. The more we come to understand about the flowers of cannabis, the more that stigma falls away. The cannabis revolution is taking hold from The United States to the EU and bringing with it a number of benefits some may not have realized. We have outlined the top eight benefits society stands to gain from cannabis legalization.

MEDICAL CANNABIS

Books could be written on the amount of information we now know about medical cannabis. Research suggests it could have implications in the treatment of cancer and chemotherapy, chronic pain, epilepsy, Parkinson's and a host of other ailments that people are currently forced to turn to pharmaceuticals for relief. Marijuana has never resulted in death from use, unlike opioid-based medicine, which causes thosands of deaths each year. As more research scrambles to investigate therapeutic potential, the movement gains new ground every day.

Numerous studies have shown that cannabinoids actively kill cancer cells. THC, the primary compound in cannabis, interacts with the cancer cell at the CB1 cannabinoid receptor. Once THC interacts with the receptor on the surface of the cancer cell, the cell releases a compound called Ceramide which breaks down the mitochondria of the cell, causing it to basically shut itself down. This effect however only occurs in cancer cells as THC does not interact with healthy cells in the same manner.

RENEWABLE RESOURCES

In addition to cannabis, legalisation is seeing the resurgence of hemp – the industrial, fibrous, non-psychoactive member of the Cannabis sativa family. Cultivated for the strong fibre, hemp was one of the first crops grown during the agricultural revolution.

It was used for centuries to manufacture rope, paper, and canvas. Hemp changed the world but since production mostly disappeared after marijuana prohibition we have all but forgotten how much of an integral part of our lives it once was. Looking forward, hemp can be used to make fuel, concrete, protein supplements, clothing as well as hundreds of more applications. It can also replace our dwindling rain forests as sources of textiles.

Hemp as a renewable resource costs less to grow than trees, can be replaced in a matter of months as opposed to years, and does not destroy fragile ecosystems in the process. The prohibition of hemp was driven by corporations who had a foothold in other industries like oil and paper. Hemp was a threat to these industries. The information age has brought an end to old myths about hemp and cannabis. As we look for greener solutions to our industrial needs, cannabis hemp reveals itself as one of the top solutions to the current environmental problems we face.

REGULATION

Legalization brings with it a need to ensure public safety, therefore, strict growing and production techniques must be implemented. This results in a better, safer product that removes any potential concerns. Government oversight guarantees that marijuana cultivated for public sale is free of toxic and potentially dangerous pesticides, and that it is grown in facilities that are as clean as can be. Regulation also brings with it product testing to determine exactly how potent your weed is.

Regulation also has the unusual side effect of decreasing use in under age consumers. Portugal decriminalised all drugs in 2001. Since their legalization, studies found that drug use, overdoses, and deaths all decreased at a steady rate. While cannabis isn't dangerous, the effects on developing minds is unknown hence the need to ensure marijuana is only consumed by those of age. Regulation has proven to be effective in curbing under age drug use and will do so with marijuana as well.

REDUCTION IN VIOLENT CRIME

Areas with legal cannabis have shown dramatic decreases in crime rates particularly violent crimes. Are people too high to fight? We have a feeling there is more to it than that. Some areas have shown dramatic decreases in reports of theft, DUI and violent crime by as much as 15%. While it is too soon to have enough definitive data, all indications show that areas with legal pot enjoy a higher quality of life with lower violence and crime rates. The studies are at least all in agreement, that there are no connections between legal pot and increases in crime.

ECONOMY

Cannabis is the highest value cash crop on the planet. It is attracting the attention of investors eager to cash in. Along with investment comes jobs. The cannabis industry is bringing in money hand over fist. With such a large consumer base worldwide, investment in a cannabis grow is guaranteed to see big bucks come through the door. Along the way, cities in the United States that have legal marijuana have seen unemployment numbers drop to historic lows. The growth of the cannabis industry has also created many other new industries that look to cater to the green industry. During a gold rush is a great time to sell picks and shovels. A world of online, media, and horticultural services have sprung into existence since the first days of legal cannabis.

The legalization of cannabis has had a significant effect on illegal drug cartels as well. Instead of money being funnelled to illegal drug organizations, the same profit is making its way into the pockets of entrepreneurs and business owners, who in turn spend that money on employees, technology, and expansion. Shifting a product from the black market to an open market takes a huge chunk of what would otherwise be illegal money and turns it into legitimate taxable earnings that will help grow our economy. Another side effect of legalization is a reduction in prescription drug costs. As the pill industry sees more people turning to cannabis, they have had no choice but to drop prices.

ELIMINATING THE BLACK MARKET

Legal cannabis drives prices so low due to basic supply and demand principles that illicit drug dealers and large drug organizations simply can't compete. Not only can they not compete financially, they cannot compete in quality. Modern cannabis businesses are hiring master growers with advanced degrees in horticulture to run their facilities. The end result is not something that you can find in an ad-hoc grow op. The war on drugs created large cartels responsible for flooding the streets with narcotics. With legalization and regulation, the black market operators can no longer thrive. Some cartels have even been busted trying to get into the legal weed game.

TAX REVENUE

Along with a booming economy comes an increase in tax revenues. Some places tax sales on marijuana as high as 40%. Legal cannabis has driven prices down. So low in fact that even after a 40% tax, legal cannabis is still cheaper than the black market. Any time the government can take a black market product and regulate it, not only do they reap the financial benefit, they make the overall experience safer. Tax revenues in places that large scale cannabis production is legal have seen new heights. Much of the new money coming in is spent on education, drug awareness, and prevention programs and medical programs for children.

Recent estimates show that European nations currently consume approximately ten billion Euros worth of cannabis each year with the heaviest concentrations being in Spain, France, Italy, Germany and the UK. The difference in cannabis use between these nations and the rest of Europe is significant, in part because of cannabis' legal status. Opening up legalization would significantly increase tax revenues in these nations. In 2015, the state of Colorado in the US generated over 125 million dollars on 1 billion dollars in cannabis sales. The increase has been steady over the years, peaking in 2015. As cannabis becomes legal, increased tax revenues is a guarantee.

NEW GENETICS MEANS NEW POSSIBILITIES

As cannabis grows in acceptance, more and more people are researching the medical potentials and breeding new strains with specific properties to treat specific ailments. The properties of the cannabis plant are self-evident, however with new strains come new discoveries. Horticultural scientists and cannabis breeders are constantly crossing varieties to find the next big thing in weed for both recreation and medical use. Crossbreeding led to the creation of the strain Charlotte's Web, the highest CBD strain and currently the most effective in fighting seizures. The list of medical breakthroughs due to cannabis research is growing every day. Breeding cannabis to push the boundaries of medicine may be the single biggest benefit that we will see as legalization continues to spread.

MEDICAL MARIJUANA A HUGE OPPORTUNITY FOR AFRICA

All African nations are capable of cultivating medicinal marijuana to exploit a growing international acceptance of cannabis-based medicines.

The current global market for marijuana products is US$3 billion and is expected to rise to US$56 billion as more countries and US states join the legalisation trend.

African nations are reluctant to get in on the business, with conservative governments fearful of encouraging recreational drug use. Cultivation of a different kind is now a reality with Africa's first license to legally deal in medicinal marijuana issued by Lesotho.

Lesotho's government has granted a local subsidiary of the South Africa firm Verve Dynamics the right to cultivate, manufacture, supply, export and transport Cannabis and Cannabis products from Lesotho.

Verve Dynamics's South Africa operations specialises in medicines manufactured from indigenous plants.

Groundbreaking work was done with Sceletium tortuosum, a succulent herb from South Africa known as Channa, Kanna or Kougoed that has been used as a tranquiliser for centuries.

Jan van Riebeeck and the crew of the Drommedaris were impressed users in 1662. Richard Davies, the managing director of the Verve Dynamics Group, holds the patent to the Sceletium tortuosum-based product Trimesemine™.

Verve has a benefit-sharing agreement with South Africa's Khoe Khoe community for Sceletium products, and intends to set up a similar community profit-sharing deal in Lesotho for the medicinal marijuana products.

Medicinal marijuana involves the use of all or part of a cannabis plant. The US Food and Drug Administration (FDA), which requires extensive studies involving 100 000 subjects to ascertain benefits and risks of new drugs, has in the absence of such studies never approved medicinal marijuana.

However, 29 of the US' 50 states as well as the capital district Washington DC do permit medicinal marijuana.

The FDA has approved two medications in pill form synthesised from cannabinoid chemicals. Cannabis is used to treat a wide range of illnesses, from childhood epilepsy to glaucoma. Marijuana reduces inflammation, nausea and pain and is being studied as a control for mental illness and addiction.

"We have been dealing with Afro-botanicals with a medicinal application for the last 15 years. Cannabis has been a product of interest to us for some time due to its medical potential. However, due to the illegality of it in South Africa we have not had access to it," said Verve Dynamic's Davies in an interview.

"Based on the work we had been doing in South Africa, we were afforded the opportunity to present our case to the Lesotho government's Minister of Health. Thereafter, we underwent an intensive six-month due diligence, part of which included a site visit to our South African facility by various departments of Government including Health, Trade and Industry, Tourism and agriculture and the Lesotho police."

With license obtained, investment began in what will be a R30 million first phase that will employ about 40 Basotho to grow and process cannabis. Good Agricultural Practices and Current Good Manufacturing Practice regulations will be followed as well as Lesotho's and international medical and narcotic regulations. Once the fields are producing a satisfactory product, they will be processed into finished products in the form of capsules, tinctures, and other medicinal consumables.

"Phase 2 and 3 would be expansion phases based on the market demand for medical cannabis. The greater the demand the bigger these phases would be but we anticipate them being at least 2 to 3 times the investment of Phase 1," said Davies.

"All staff, as far as possible, will be sourced from Lesotho. Initially some of the manufacturing staff will need to move over from South Africa until we have successfully completed the necessary skills transfer," he added.

As new markets open overseas, demand will grow along with profit potential. "We will be able to sell product to any country that is legally allowed to buy the product. Our main markets are expected to be North American and the European Union. I'm of the opinion that the revenue potential for Lesotho, as well as the rest of Africa is obvious. Having the ability to deliver a superior product to the international market at a lower than market related price is a really good position to be in," said Davies.

Lesotho seems to have nothing to lose from the venture, and as the first African country out of the gate in an expanding global business will enjoy the benefits usually reaped by pioneers. Beyond Lesotho, Davies said, "We are of the view that all African countries could benefit from the legalisation of medical Cannabis. Africa, particularly Southern Africa, is rich in the resources required to propagate and produce medicinal plants."

Lesotho's sister kingdom, Swaziland, is noted for high-jgrade marijuana, and theoretically a firm may be licensed to produce the otherwise illegal plant under the country's Pharmacy Act. Royal Swaziland Police Force spokesman Assistant Superintendant Khulani Mamba said in an interview, "It has never happened but someone could seek a license from the Ministry of Health. A board sits to consider applications for producing new medicines in the country."

Other than that, marijuana is outlawed in Swaziland, and police are obliged to follow current laws on recreational drug use by arresting possessors and destroying marijuana fields.

While marijuana has potential for other industrial uses – clothing, lotions, lubricants, paper, ropes and even bricks and bird seed have been made from hemp/cannabis – worries about recreational marijuana arriving through the Trojan Horse of medicinal marijuana can only be alleviated through stringent regulation, Davies believes.

"When dealing with a consumer good, particularly one that is used for medical treatment, a level of regulation is necessary for the safety of the consumer. There is an ethical responsibility to provide consumers with a predictable and reliable product that is safe and effective. Alcohol is perhaps a good local example of this. You cannot just make a home batch of alcohol and sell it to others. These regulations are for public benefit and safety," he says.

"Medicinal cannabis can be thought of as the gold rush of our time. Africa can be at the forefront of this industry," Davies believes.

SWAZI GOLD

"Swazi Gold" is a pure sativa strain that is said to take six months to mature and can grow over ten feet in height. Additionally the plants are rumored to be completely devoid of the cannabinoid CBD, but Hamilton wanted to test this for himself. So he began collecting samples... for "analysis." Contrary to the popular image of cannabis growers as stoners and hippies, the growers of Swaziland are predominantly drug naive grandmothers who grow the plant out of pure economic necessity.

With the country in fiscal crisis and its former wood pulp and mining industries exhausting the last of

Swaziland's natural resources, Doctor Ben Dlamini has a few remaining solutions to attract foreign investors and protect the country from economic collapse. But the market is no longer dominated by the original six-months Swazi Gold and there's now a demand for fast-growing hybrids imported from Europe.

Africa produces more cannabis than any other continent in the world and Swaziland, despite its small size, dedicates more hectares of land to cannabis cultivation than all of India. One interesting thing about cannabis growers in Swaziland is they never get high on their own supply. Farmers report, the iron rich soil is ideal for growing cannabis without the use of synthetic fertilizers.

THE MARIJUANA BUSINESS IS REALLY THE REAL ESTATE BUSINESS

Selling weed seems like a cash cow, but the real money these days is in the real estate that represents the most crucial part of the cannabis business.

In the unstable and risky marijuana industry anything can happen--your neighbor could wage a Racketeer Influenced and Corrupt Organizations Act suit against you, your bank can tell you to move your money to another institution, or you could lose your license for a small screwup. But in all of this uncertainty and risk, pot entrepreneurs are making one safe bet: buying real estate.

"With so many obstacles and regulations in our way, owning your real estate is the only thing we can control in this industry," says Sally Vander Veer, co-founder and CFO of Denver-based marijuana cultivator and retailer Medicine Man. "It's essential to long-term success."

Aside from being a simple and time-tested investment, Vander Veer says owning your own real estate is also a smart way to safely store your money, as long as property values hold or increase. And surprisingly, property ownership is actually a flack jacket that protects your business from a frequent nightmare scenario in the pot business.

Since 2013, many cannabis entrepreneurs have seen their warehouse rents skyrocket after they'd spend tens of thousands of dollars to convert the space into a marijuana growing operation. Owning your real estate lets you avoid that financially crippling scenario entirely, says Patti Zanin, an independent real estate agent in Denver who serves weed clients.

Zanin says buildings zoned as "light industrial" that have been vacant for years are now the most valuable properties in the area thanks to the marijuana industry. Properties go so quickly, Zanin says, that a secondary market has blossomed--well-heeled companies will buy a property, get a license, and sell the whole package to smaller businesses.

After spending over $1 million to build out the 40,000-square-foot warehouse Medicine Man rented on Nome Street near the Denver airport, Vander Veer and her brothers Andy and Pete Williams decided to buy it. The brothers and sister got it for $2.5 million in 2014 and it's now worth $6 million (they recently sold it).

"It's the green boom here in Colorado and real estate is at a premium," Vander Veer says. "If you own a building that is zoned properly, not including any improvements, it's worth millions."

Medicine Man just bought a property in Commerce City to build a new dispensary and rent out retail space to a few non-marijuana tenants. "Part of our overall strategy is to own all of the real estate we are in. We're looking toward a hard asset mentality," she says.

The property is also a form of banking in an industry that still has trouble maintaining bank accounts. Bankers need to do more compliance work for marijuana company customers, which makes the accounts expensive--some cannabis entrepreneurs pay $3,000 a month in account fees. Since it's expensive to bank cannabis profits and banks won't loan to cannabis entrepreneurs because marijuana is still federally illegal, buying real estate with cash profits is an increasingly popular alternative.

"You have to find a safe place to store your money," Vander Veer says. "Where else but real estate?"

It's not the weed business.

Josh Ginsberg, co-founder and CEO of Native Roots, says his company has gotten so large that he has a team tasked solely with looking for property to buy or lease. Native Roots has 450 total employees. Ginsberg says the real estate team has been key to the company's explosive growth in the past two years, from four to 15 stores. The largest building Native Roots owns is a 180,000-square-foot grow facility with more than a hundred employees and tens of thousands of plants. Shortly, the company will open its 16th location, a third gas station-pot dispensary in Colorado.

The company owns half of its buildings and is working on buying more. Ginsberg says as a marijuana company that is growing as fast it can, he does not want to be at the mercy of anyone, especially a landlord.

"We put a lot of money into our stores to make them how we want, and it doesn't make sense to put this much money into someone else's building," Ginsberg says.

Meet the landlords.

The Kalyx, a real estate investment trust, is trying to be the marijuana industry's "honest landlord." The company, which is based in New York, was founded by Potter Polk, a tourism and internet entrepreneur, and George Stone, a founding partner at real estate investment firm The Witkoff Group (which bought the Woolworth Building in New York City, among other properties).

The company raised $25 million to buy properties in states where marijuana is legal, including three in Denver, one in Auburn, Washington, one in Eugene, Oregon, and a sixth in Phoenix, Arizona. Polk and Stone then rent the properties to marijuana businesses with the agreement that rent will increase only 3 percent a year. In an industry where rent can increase by 50 percent from year to year, Kalyx's business model is enticing to marijuana entrepreneurs.

"Because the industry is so small, you want to be the guy who is fair. We're building our business along side with these guys, not at their expense," Stone says.

Kalyx owns buildings in Denver that are leased to three big weed brands--Strainwise, Medicine Man, and Dixie Elixirs.

The seller's market.

Rona Hanson, a real estate agent who finds homes

that accommodate home grows, says the cannabis real estate boom is having a knock-on effect on the residential market. Home values have gone up 13 percent since last year, she says, and a house under $300,000 will typically have 30 to 40 offers after an open house and on average sell for 103 percent of asking price.

Some cannabis entrepreneurs are even pivoting away from the retail pot business to pursue real estate riches while the getting is good. Aaron Herzberg, partner and general counsel at marijuana holding company CalCann Holdings, owns three dispensaries in California (including actress Roseanne Barr's pot shop). He says CalCann will start to focus on real estate instead of running dispensaries because marijuana properties are a hot commodity--a $2 million piece of land could go for as much as $10 million if zoned properly, he says.

They are looking to raise $10 million in the next three to four months to staff up and start buying more weed-friendly properties. Eventually, they want to become a marijuana investment trust and buy up properties in newly legalized markets.

Herzberg says his company worked with the city government of Adelanto, California to pass a city ordinance to allow for cultivation and in the process bought seven parcels of land for $350,000 before the public knew about the pot law. Two months later, they sold the land for $1.9 million to hopeful marijuana entrepreneurs.

"After going through that deal, we came to the realization that running dispensaries is a great business, but it's a lot of work, it's not all that profitable and it's challenging to work with the regulations," Herzberg says. "Doing the real estate, while it is not easy, is just a better business."

AFRICA, THE NEXT BIG CANNABIS MARKET?

Canada is leading the way in the cannabis industry. With marijuana legalization happening being voted on this summer, Canada is embracing the business of pot. As the industry grows, however, international markets are expanding across the Atlantic to the continent of Africa.

The Globe and Mail report that Supreme Cannabis Co. (TSX-V:FIRE) "has invested $10 million in a Lesotho company, with the aim of exporting high-quality cannabis oil to Canada and other markets." Other Canadian companies are also eyeing Africa as a cannabis partner. The same article reports that "Canadian investors are scouting for 10,000 hectares of land for marijuana farming in the Mashonaland region" of Zimbabwe.

Lesotho Issues First Manufacturing License

Lesotho became the first country on the continent to issue a medical marijuana license to Verve Dynamics. The South African based company specializes in botanical extracts. Lesotho is an attractive country for companies like Verve because of the black-market cannabis industry that already exists. According to Quartz Africa, "Lesotho's farmers have already been growing weed for the consumption at home and across the border in South Africa." In fact, the black market has helped many of the poor farmers there generate income.

Like other countries entering the cannabis industry, Lesotho is eyeing the possible financial windfall that could accompany the crop —something which is sorely needed in a country with a high poverty rate. Other African nations are contemplating legalization as well. According to Quartz Africa Malawi is taking the same approach to cannabis as Lesotho. They too have a thriving black market and view legalization as a path for the country to generate additional revenue.

South Africa Opens Africa's First Dispensary

One African nation has gone one step further with the marijuana industry. Earlier this year, South Africa opened its first dispensary. Canna Culture is a Durban-based dispensary that features Dr. Kyle Deutsch, musician, and former Idol SA contestant, as the in-house chiropractor.

The dispensary, which features medical cannabis products from the United States was developed to ensure that South Africans were getting high-quality weed to treat their ailments instead of black market products that were not regulated. Founder, Krithi Thaver, told Time Live, "What I realized in the last year, there are so many people becoming kitchen pharmacists where they are going online, checking a formula on how to make cannabis oil and making money off it without actually curing a patient."

While the company is focused on the medical side of the industry, they are also preparing for full legalization when the time comes.

If you enjoyed this book would you do me the honor of leaving a review on the Amazon page. Your support is very much appreciated. Thank you and God bless you.

www.ingramcontent.com/pod-product-compliance
Lightning Source LLC
Chambersburg PA
CBHW071600220526
45469CB00003B/1072